Asymptote Einstein

Elements & Environments

By

Damon Dion Reed

Table of Contents

Chapter 1: Idiots 3

Chapter 2: Hinterland 6

Chapter 3: Gross.. 17

Chapter 4: Revival... 22

Chapter 5: Density 27

Chapter 6: Hypothesis at Work 30

Chapter 7: Salty... 34

Chapter 8: Justified 39

Chapter 9: Esoteric.. 43

Chapter 10: Stimulated Tooth Emission 49

Chapter 11: Einsteinium 52

Chapter 12: Cold Dinosaurs 61

Chapter 13: Garrulous 66

Chapter 1: Idiots

Einstein was an idiot. He used cocaine, had affairs, and made mistakes. Now the reason why I'm apt to inform you per Einstein's idiocy is because intelligent people are the ENEMY! We hate it when smart people make us feel incompetent. We need to feel good about ourselves and that is the reason why we let stupid people run our country. Or in simpler terms, our comfort is based upon being able to insult stupid people such that we can drink a beer while watching the game and/or fall asleep without a care in the world. That is why Einstein was an idiot. In fact, he is such an idiot, he was MAN of the twentieth century...or so People Magazine tells us. (Personally, if you've got a beer belly and a rusted car in the front yard, then you're more of a man then Einstein could ever be.) Now don't get me wrong, Einstein wasn't the ONLY idiot on this planet: Bill Gates is an idiot; Stevie Wonder is an idiot; Richard Feynman was an idiot. In fact, everybody with an advanced college degree or a ground breaking talent is a COMPLETE and utter idiot! So sit back and relax because America was built on complacency.

Having this in mind, Freud was the biggest idiot because he gave 'names' to the reasons why people are predisposed to reject authority, laugh at another's misfortune, and self-aggrandize. Hey Freud! We don't need words like Ego, Id, or Superego to understand why there's an ANTI-intellectual movement in America. America was built on complacency, so shut the fuck up Freud!

Also, just because you went to an Ivy League school and/or you've got a rich parent, doesn't mean you stand above everybody else. Tons of idiots have gone to Ivy League schools. But, if you rationalize that nothing good can come from going to a two-year technical school, then you're totally missing the point, i.e. you've got a problem with one of those stupid Freud-words. Either that or you're just another shmuck that thinks that the educational system is broken, which means YOU shouldn't try because you'll never achieve anything anyway.

Whatever the case may be, Einstein was a complete idiot, you rock, and who cares about Gaussian Curves and Asymptotes. You're here, you're not queer, and you've got the right to change the channel any time. So go ahead. Change the channel to something that doesn't make your head hurt. In fact, I DARE you to watch something with absolutely NO moral, civil, or educational repercussions. (FYI, I know you want to change the channel.) You've worked hard all day and you deserve to feel good about yourself. Let somebody else educate your children and/or fix the World's problems. You just sit back, feel that ice cold beer in your hand(s), and bask in your own, magnificent, glory.

With that delightfully psychological introduction out of the way, let me be the first to introduce you to the future. Hiya and how ya doing? In any event, I could be a complete egotistical ass or I could be working on the first FUNCTIONAL unified theory in the history of humanity. (You get to decide.) Butt, no matter how **functional** my unified theory may or may not be, it will ALWAYS be a working hypothesis because nobody is perfect, we can't see everything, and human life is much more finite than the universe. Or in simpler terms, this book is just another attempt at improving my unified theory, i.e. Tiny Bang-Bang^{Asymptote} Theory. And on the off chance this book doesn't render a perfect unified theory, then I'll write another book. (Most definitely the case.) But trust me when I say the following: My stepwise de-stupidification is a stepwise process, i.e. buy ALL my books...please?

Chapter 2: Hinterland

As great as humanity proclaims to be, there is still room for improvement. Unfortunately, 'improvement' is a relative term. How about this instead: I want to go to space. (That was a selfish non-relative statement that most people can relate to.) More specifically, I want to go to the Moon. And even more specifically than that, I want to poop on the Moon. (One small turd for me, one giant turd to confound aliens.) Butt, before I get to do all that, space travel needs to be more prevalent in society.

The problem with getting to space is pushing through ALL the 'stuff' that is TRAPPED in Earth's gravitational cage, including Earth's gravitational cage. Or in more specific terms, spaceships have to push through tons of atmosphere, plasma, negative magnetic energetic quanta, and away from Earth's negatively diffusive core. All of which means, no matter how much money America throws into space exploration, it will NEVER get far using OLD-fashion HOT-BUTT rocketry. Therefore, instead of using insane amounts of combustive fuel, and the negative plasma it releases upon combustion, to PUSH a spaceship away from the Earth's **equator**, we should launch space exploration 'devices' near Earth's poles. Or in simpler terms, the atmosphere is thinner near Earth's poles, which means there is less plasma/heat to push through. Or in scientific terms, less negative thermal energetic quanta within AIR's atomic orbitals will result in LESS 'mass' to push through. (Launching at night will also be more energy efficient and fun because the saucers at the bars will think they're seeing flying saucers!) Also, there will be less resistance moving through cold-dense atoms because the atomic orbitals are less likely to be distorted, which causes the release of negative thermal energetic quanta. And while we're on the topic of negative thermal energetic quanta, here is something that just came to mind.

Do you remember Anisotropy? Well, it the transduction of proton 'spin' energy through space by negative magnetic energetic quanta, which is the result of negative thermal energetic quanta causing positive protons to degrade. All of which means, the release of negative thermal energetic quanta can cause the attraction between objects via the release of OPPOSITE negative magnetic energetic quanta. Or in extremely shocking terms, GRAVITY is a function or movement. On second thought, let me try and explain that a little better.

I just went to Home Depot to buy a lock for my gym locker. While I was there, I saw Fredrick, who is a co-worker of mine. And for the briefest of seconds, I thought Fredrick was going to extend his hand to greet me, which would have **slowed** my desired in-&-out shopping experience. (FYI, I wasn't working and I like Fredrick.) And this is exactly the WEIRD component of gravity that has influenced NASA to cover their rockets in NON-protonic material. Unfortunately, electrons can undergo stimulated degradation by negative thermal energetic quanta to release negative magnetic energetic quanta, which CREATES a 'gravity' factor. Or in the simplest terms, the faster an object moves through the air, the more that object will distort atomic orbitals in the air, which will result in the release of negative magnetic energetic quanta by the air molecules. (FYI, this is dependent on the temperature of the air.) In addition to the air molecules releasing negative magnetic energetic quanta, the atoms in the airplane will do the same. And as a result of OPPOSITE negative magnetic energetic quanta being attracted to each other, the **faster** the 'object' moves through the air, the more the object will be ATTRACTED to the air.

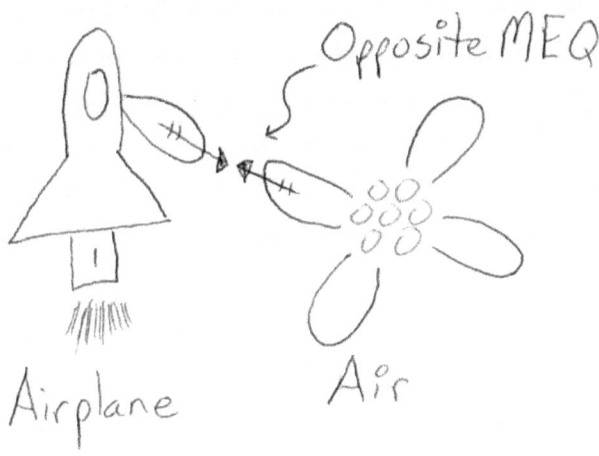

Figure 1: Drag Queens

As you can see in this disturbing figure, I've gone against the laws of physics and exaggerated the **size** of an AIR atom and the **size** of an ATOMC ORBITAL on an airplane such that you can see how the movement on one relative to the other, and the ensuing atomic orbital distortion, will result in OPPOSITE magnetic energetic quanta via stimulated degradation, which are attracted to each other. (Long sentence.) Or in the simplest terms, I just explained 'drag' in terms of *Quanta Dynamics*, which is a factor of 'gravity' and speed. Now, back to exploring the universe...on a budget.

In addition to all the above mentioned benefits of launching rocks from Earth's COLD poles, before I went off on a tangent about Quanta Dynamic Drag, Earth's poles have a predominance of ONE negative magnetic energetic quanta, which could be beneficial to space exploration. Or in complex scientific terms, you know that thing that superconducting COLD magnets can do, i.e. float? Well, you could create a spacecraft that could do the same thing. Or in the simplest terms, check out the ensuing figure.

Figure 2: Spaceship 101

As you can see in this figure, Mr. Spaceship is using the REPULSIVE power of SIMILAR magnetic energetic quanta to float up-up-and-away from Mother Earth. Or in even simpler terms, the north pole of the spaceship's magnet REPELLS the north pole of Earth, i.e. the North Pole. Unfortunately, since humanity has NOT discovered a method to SELECTIVELY degrade electrons to produce MONO-pole magnetic entities, the spaceship will have to use the South-pole magnetic energetic quanta within the spaceship to create a plasma battery. (If only someone invented a chiral nanotechnology coating that released magnetic energetic quanta when heated by air friction, then air planes would be more buoyant.) In any event, using magnetic repulsion between the spaceship and Earth will decrease the weight of the spaceship.

Another way to decrease the weight of spaceships is by using 'Earth based magnetic cannons' to **fling** spaceships into the atmosphere. First and foremost, this will lessen the weight of the spaceship by using Earth based magnetic acceleration. Next, Alaska has lots of land to build long tracks of covered magnetic acceleration tracks, which will also alleviate the concerns of Alaska's harsh weather building up on spaceships that are waiting to launch. And finally, Alaska is COLD, which is a good because super-conducting MAGNETS need to be kept COLD. Or in simpler terms, it won't cost as much to COOL the Earth based electromagnetic propulsion system in Alaska.

With all that verbal fluff in mind, with regards to spaceships, I think it is a perfect time to introduce some reality into chapter. More specifically, anti-gravity. And even though I've talked about this before, I've come up with an even more interesting way to describe it. But first, let's render Earth's magnetosphere to your imagination.

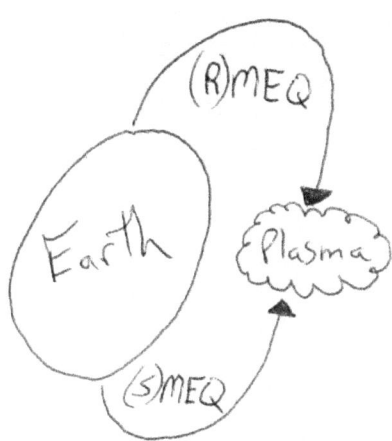

Figure 3: Earth's Magnetosphere

As you can see in this figure, the PERFECT collision of OPPOSITE magnetic energetic quanta about Earth's equator produces plasma. Unfortunately, since Earth has an atmosphere with tons of stuff in it, the perfect collision of these OPPOSITE magnetic energetic quanta **exactly** at the equator is statistically low. Or in more specific terms, please look down.

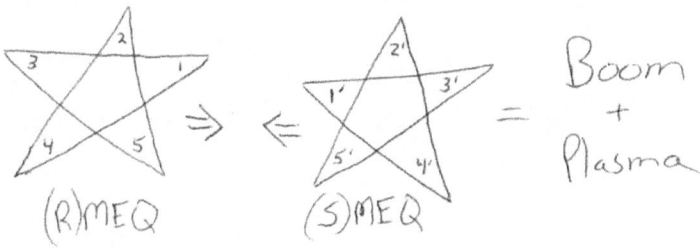

Figure 4: The Perfect Collision

As you can see in this figure, the PERFECT collision between OPPOSITE magnetic energetic quanta results in BOOM + plasma. Granted, you can super accelerate energy and make it go BOOM regardless of its alignment, but that is NOT the case in Earth's magnetosphere. What is the case in Earth's magnetosphere is as follows:

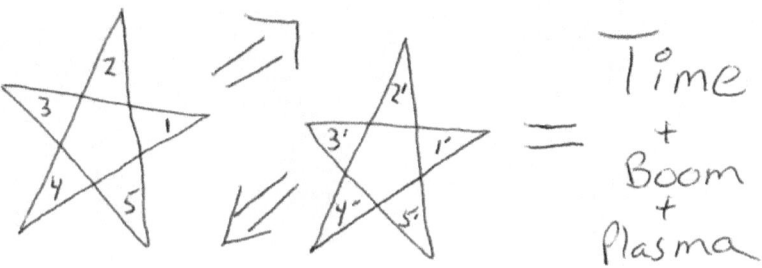

Figure 5: The Imperfect NON-Collision

As a result of the imperfect NON-collision of these opposite magnetic energetic quanta, they DO NOT collide, which means they are given more TIME to exist before they go BOOM + plasma. Or in the simpler terms, even though the equator has an abundance of PERFECT collisions between opposite magnetic energetic plasma to result in a warmer climate around Earth's equator, Earth has a large MIXING region of OPPOSITE magnetic energetic quanta. Or in the simplest terms, Earth's magnetosphere looks more like the figure below.

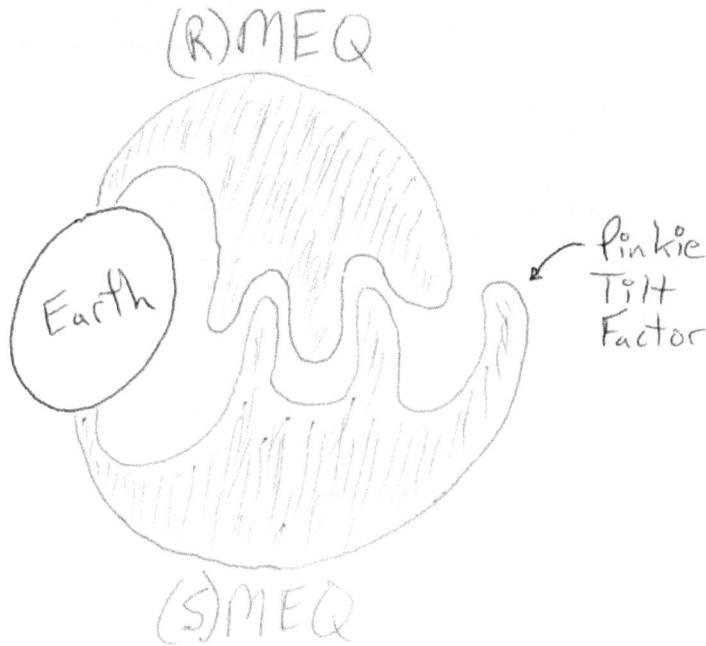

Figure 6: God's Hands.

Hopefully you can imagine, based upon the NON-perfect trajectory of OPPOSITE magnetic energetic quanta in Earth's magnetosphere, a TON of (S) MEQ exist in the northern hemisphere and a TON of (R) MEQ exist in the southern hemisphere. But, the closer the magnetic energetic quanta get to the opposite pole, the less likely they are able to AVOID a PERFECT collision and go BOOM + plasma. All of which, leads us to the PINKIE TILT FACTOR (PTF).

As you can see in the previous figure, God has an OUTER pinkie that is further away from Earth. Or in scientific terms, based upon the IMPERFECT nutshell we call Earth, ONE magnetic energetic quanta is more prevalent in Earth's magnetosphere. Or in other terms, based on the degradation/diffusion of ONE magnetic energetic quanta with regards to Earth's IMPERFECT crust, Earth's magnetosphere has an abundance of ONE magnetic energetic quanta, i.e. the PINKIE TILT FACTOR.

Now as uninteresting as the Pinkie Tilt Factor might sound (#penis), the Pinkie Tilt Factor determines the TILT of Earth towards the Sun. And for those of you who have been paying attention, this is a CORRECTION with regards to my postulate about a MASSIVE magnetic energetic quanta spectrum, which creates a solar magnetosphere. Now don't get me wrong, there is a SPECTRUM of magnetic energetic quanta, but is probably NOT like what I described it in *Thinking Outside My Gated Community*. Unfortunately, to get the whole gist of the Pinkie Tilt Factor, I need to lubricate the Pinkie Tilt Factor with a little *Quanta Dynamics*.

We exist in a negative branch of the universe, which causes positive matter to be larger. Having said that though, electrons still contain a considerable amount of positive energy to swage the columbic repulsion of all the negativity energy within the negative electrons. And even more importantly, negative magnetic energetic quanta contain a greater amount of positive energy components in comparison to light, which is the reason why magnetic energetic quanta explode in deep-dark-COLD positive space between galaxies. Therefore, the Pinkie Tilt Factor is SIMPLY a function of GREATER positive **energy** being released by ONE of Earth's poles, which is attracted to the Sun's constant negative diffusion. Or in simpler terms, the **relative** positivity released by ONE of Earth's magnetic poles causes Earth to tilt towards the negative Sun.

With all that humiliation out of the way, the trick to creating an anti-gravity device is to create a blanket of magnetic energetic quanta. And just in case you were wondering, it should be an interwoven blanket of BOTH magnetic energetic quanta. (FYI, spaceships launching from Earth's poles need to maximize thrust. So they NEED only **one** magnetic energetic quanta.) Or in simpler terms, look down.

Figure 7: The Blanket

Hopefully you can surmise, based upon the **trajectory** of the opposite magnetic energetic quanta, there will be LESS perfect collisions/degradations about the magnetic energetic quanta being released from the anti-gravity device, which means there will be MORE magnetic energetic quanta to deflect/degrade Earth's magnetic energetic quanta. As for the technical minutia with regards to the creation and release of these anti-gravity magnetic energetic quanta, I'll save those postulates for later. Maybe spinning magnets or alternating magnetically degrading atoms...who knows?

In conclusion, I want to poop on the Moon and I won't stop until this dream is a reality! Also, Earth's imperfect crust has created a slanted relatively-'positive' magnetosphere, which tilts Earth towards the negative Sun, i.e. Pinkie Tilt Factor. Or in religious terms, hot chicks always like a little stinky pinkie after you lube it up with some tongue juice? (FYI, the 'shocker' is when you stick two in the poo or say 'in religious terms' in front of a highly sexual statement.) And finally, focusing our resources on energy efficient space travel, based upon *Quanta Dynamics*, might led us to building a massive spaceport in Alaska.

Chapter 3: Gross

I know I shouldn't be doing anything else when I'm driving, but I live in the middle of nowhere. As a result of a half-hour drive to 'civilization', I floss and brush my teeth when I'm driving. And since I don't have a sink, I use a glass filled with mouth wash to clean my toothbrush. (Totally gross and yes, I don't date.) In any event, the other day when I dipped my toothbrush in the mouthwash and tried to shake the toothbrush clean, my brain said 'abnormal-mass'. Fortunately, as a result of concentrating on this topic for several years, my brain didn't stop there.

Thus far, I've talked about 'abnormal-mass' as it pertains to a spectrum of volume variations based upon the environment and the attraction between relatively DIFFERENT negative objects...in our negative branch of the universe. (I also talked about abnormal-mass as it pertains to stars, black holes, second matter, second stars, and second galaxies, which are second black holes, but all that has nothing to do with this chapter...I think.) In addition to all that, 'abnormal-mass' can be the result of plasma diffusion INTO porous matter, which is related to the TYPE of environment. Have said that though, I need to take a huge step back and unequivocally state the following: Matter is created by the arrangement of complex energy. In simpler terms, the OLDER matter becomes, the more porous it becomes, and the more plasma it can absorb. (#SaggyProtons) All of which, complicates the understanding of gravity about old galaxies and second black holes, but that is also beside the point...I think.

Even though the IDEA of porous matter containing 'Energy Containment Regions', much like atomic orbitals, is somewhat disturbing, it does open up a method to understanding environments and possibly, different branches of the universe. Unfortunately, this is the point when mathematicians blow their tops because we have no way of knowing if the stimulated degradation of matter by external energy consumes intercalated plasma or causes the release of intercalated plasma. But one thing is for sure: Intercalated plasma CANNOT spontaneously forge more complex energetic quanta like photons or magnetic energetic quanta because intercalated plasma has been tainted by decay and expansion, which limits its momentum.

With that SPECIFIC criteria in mind, there is an abundance of questions that jump to mind, or more specifically, jump to my mind. For starters, is the ABSENCE of intercalated plasma the major factor in proton degradation in deep-dark-COLD positive space? (Answer: We are in negative branch of universe, which means intercalated plasma is probably negative, i.e. YES.) Next, how does intercalated plasma stabilize the electronic structure of protons? (Answer: Picture on the front of this book?) Also, is the TYPE of residual plasma the **factor** that determines the TYPE of STABLE energetic quanta, i.e. matter? (Answer:

My head just exploded as to a METHOD.) And finally, how does all this relate to stars? (Answer: In next paragraph.)

I don't believe that in a DYNAMIC universe, there are just THREE stable energetic quanta. Or in simpler terms, matter is a function of environment. Or in complex terms, there is a spectrum of star births, which is directly related to the environment and the matter produced. In as much as the preceding statement is slightly confusing, let me try and clarify it with a figure.

Figure 8: Baby Star Spectrum

Previously, I hypothesized that matter is popped like popcorn around stars. But, maybe this figure is easier to understand. If a Baby Star is born in deep-dark-COLD positive **field** without any shelter, then it won't produce as much matter. If a Baby Star is born in a semi-structured **shack**, then it will produce some matter. And finally, if a Baby Star is born in a **community**, i.e. a galaxy, then it will produce all sorts of matter. All of which, is based upon the ENVIRONMENT that supports the STRUCTURE of energy. Or in scientific terms: Stars born in a relatively positive environment will diffuse copious amounts of negative energy before producing protons; Stars born in a semi-negative environment will produce more protons; Stars born with a negative spoon up their ass will produce the MOST protons...theoretically.

Unfortunately, all of this obtuse thinking now requires me to bring up the following topic: Our existence is based upon a universe forged by degradation. Or in simpler terms, even though I've mentioned that positive protons are popped like popcorn about negative stars, this question remains: Is the stability of the average proton volume based upon degradation? Or in the simplest terms, does a proton have to be bathed in continual star negativity, which causes it to degrade, such that reaches a maximum longevity with regards to its average volume within an average negative environment...in our negative branch of the universe? I know that it is tough to imagine that protons REQUIRE continual degradation to maximize their longevity with regards to a particular environment, but just imagine that baby protons are JAGGED rocks that NEED to be smoothed out by a star's continual flow of negativity. Or in scientific terms, when protons are born, they **don't** AUTOMATICALLY expand to a given size because their volume is based upon environment and their internal electronics. Also, based upon energy conservation and the fact that protons bask in a star's negative glow for BILLIONS of years before forging heavier elements, it would seem logical that protons are born with just a **microscopic** excess of positivity, which is exaggerated by degradation over BILLIONS of years to form the protons, and elements, we know and hopefully, love. All of which, reinforces the concept of proton production based upon a star's environment.

In conclusion, in as much as volume vibration is a function of the presence, absence, and TYPE of negative thermal energetic quanta with regards to the quantum fragment in this negative branch of the universe, it should be apparent how different fragments of God's pet-singularity probably contain unique energy fragments, i.e. 'matter', based upon the smallest fragments of energy. And even though we may never see it, based upon the extrapolation of *Quanta Dynamics*, learned from this branch of the universe, there is more than likely a universe full of unique matter particles, which are forged by electronic structure and electronic environment!

Chapter 4: Revival

In as much as matter is a function of negative plasma stabilizing the structure of matter in this negative branch of the universe, the structure of the matter determines the effect of gravity on the matter. Or in unified theory terms, particles and planets are similar. Unfortunately, even though these principles are unified, particles do NOT look like **little** planets.

As described in the last chapter, the presence of negative plasma about the complex electronic structure of matter selects for the type of matter that is stable in a branch of the universe. Or in simpler terms, look down.

Figure 9: Now look up.

This figure is a very VERY simplistic representation of matter in which the lines represent intercalated FAST moving ENERGY and the dots/dashes represent SLOW moving plasma. Therefore, the trajectory and the movement of the lines are directly proportional to the amount and type of dots. Or in scientific terms, negative plasma provides structural support and a protective coat to the FAST moving ENERGY within matter. All of which means, the negative plasma plays an INTEGRAL part in HOW the matter reacts to gravity. Fortunately, when we extrapolate this concept to planets, it gets a little easier to imagine.

Earth is surrounded by a negative magnetosphere, which is created by negative magnetic energetic quanta and plasma. All of which means, the component of gravity that is based upon the attraction of Earth's positive elements towards external negative energy is DEPENDENT on Earth's negative magnetosphere. Or in simpler terms, external negative gravitational energy must PUSH through Earth's negative magnetosphere to innervate with Earth's positive elements, which will PUSH towards the negative gravitational energy via atomic orbital modulation. Therefore, Earth's magnetosphere, which is created by negative magnetic energetic quanta and plasma, is SIMILAR to the negative plasma that enshrouds and intercalates the complex energy in matter.

With those two juxtaposed postulates in mind, it should be easy to imagine that as Earth's core grows more positive and releases less negative magnetic energetic quanta, i.e. a weaker negative magnetosphere, Earth will be MORE attracted to negative energy, be it from the Sun or the reflective Moon. All of which means, the evolutionary variance with regards to Earth's age and its position in our solar system, with all other factors held constant, will be **based** upon the ELEMENTS within Earth. And since Earth has a thick layer of water, which contains two positive protons, let's take a moment and imagine Earth without a negative magnetosphere.

Without a negative magnetosphere to keep negative plasma from diffusing, Earth's water will freeze. And even though Earth's frozen water will eventually lyophilize, let's imagine there is a period when Earth still has a ton of frozen water. During this period, the NEGATIVE light reflected by the Moon will result in Earth's POSITIVE protons pushing towards the 'negative' Moon, which means the Moon will move closer to Earth.

Now to the average reader, it would seem as if the Moon will eventually collide with the Earth. But, based upon the asymptotic behavior of Earth's core with regards to the elements there in, Earth's magnetosphere varies, which means the Moon's movement will be based upon the Earth's magnetosphere. Or in chemistry terms, imagine there is a chemical bond between the Earth and the Moon, which vibrates over the course of millions of years based upon the Sun's negative output, the Moon's reflectiveness, the strength of Earth's magnetosphere, the elements on Earth, and the state of the elements on Earth. Now, with that cognition in mind, quickly juxtapose this complexity to the quanta world.

In the quanta world, asymptotic events occur when part of matter's complex energy is destroyed, which results in the release of an energetic quanta and another asymptotic event. For example, let's imagine a POSITIVE proton. When negative plasma destroys part of the positive proton, the proton undergoes an asymptotic event and becomes relatively negative. As a result of the columbic repulsion inherent within the proton's complex energy structure, the protons undergoes another asymptotic event to release a NEGATIVE energetic quanta, which makes the proton relatively positive again. All of which, is based upon the type and amount of negative plasma 'coating' the positive proton. Or in simpler terms, how a proton reacts to the exterior world is based upon the CAGE of negative plasma around the proton, which is similar to Earth's CAGE of negative magnetic energetic quanta and plasma.

In conclusion, the asymptotic events within the quanta world and about our world, Earth, dictate the behavior of the objects per our understanding of physics. And even though the occurrence of asymptotic events within the quanta world and about planets are on MASSIVELY different time scales, there are similarities between the two. Within the quanta world, the type and amount of negative plasma that supports the FAST moving electronic structure determines the volume vibration of the matter, which modulates how it interacts with the universe. About Earth, the strength of Earth's negative magnetosphere determines the attraction between Earth's positive elements and the negative energy reflecting from the Moon, which means the Moon may not crash into us after all. (Thank God!) I guess it all depends on which direction the Moon is vibrating when the shit hits the proverbial fan...in our solar system.

Chapter 5: Density

As odd as this might sound, the density of water is a function of the SMALLEST and MOST abundant energy fragment in our negative branch of the universe. Or in simpler terms, God planned it that way. Or in other terms, water is densest at 4° Celsius, which has prevented the oceans from completely freezing. But how and why?

For those of you don't remember what color you are, just look in the mirror. And when you look in the mirror, remember that atomic orbitals REFLECT certain types of energy and DEGRADE certain types of energy based upon energy conservation. Or in simpler terms, the density of water is a function of energy conservation. But before we get to all that, here are a few things you might want to remember:

1. WE live in a negative branch of the universe.
2. Negative energy causes positive energy to degrade.
3. Atomic orbitals like specific types of energy in order to conserve energy.

With that in mind, here is my working hypothesis, via Quanta Dynamics, as to how and why water is denser at 4° Celsius...in figurative terms.

4°Celsius

0°Celsius

Figure 10: Density

Since HEAT is the presence of negative thermal energetic quanta, which causes the degradation of positive protons, the protons in water at 4° Celsius degrade and release magnetic energetic quanta that degrade in oxygen's anti-bonding atomic orbital, which stabilizes SMALLER anti-bonding atomic orbitals. (FYI, the arrow with two hash marks signifies the protons releasing magnetic energetic quanta.) But, at 0° Celsius, there is LESS negative magnetic energetic quanta, which causes the protons to degrade LESS magnetic energetic quanta. And since there is LESS magnetic energetic quanta to degrade into thermal energetic quanta to stabilize oxygen's smaller anti-bonding atomic orbital, the anti-bonding atomic orbitals expand to make FROZEN ice LESS dense.

In conclusion, LESS negative thermal energetic quanta within water means the LESS positive proton degradation and the release of negative magnetic energetic quanta, which degrades into thermal energetic quanta to stabilize oxygen's smaller anti-bonding atomic orbitals. All of which, is a function of the SMALLEST and MOST abundant negative energy fragment, which is dependent on the environment created WITHIN this negative branch of the universe.

Chapter 6: Hypothesis at Work

When rendering a working theoretical environment within your head, it is best to find a focal point. The focal point of Dreamstar was the exchange of a positron between protons and neutrons within atomic nuclei. Mostly because it had already been postulated, i.e. the LOSS of a positron NOT the SHARING of a positron. But truth be told, my initial postulate was that a NEGATIVE quanta was being exchanged between Neuproz pairs. I don't know why that was my first postulate, but I adjusted my theoretical comprehension to the exchange of a 'positron' and it worked rather well. BUTT, as further cognition rendered a clearer view of the theoretical environment within my head, it became apparent that MORE THAN likely my first postulate, i.e. the exchange of NEGATIVE quanta between Neuproz pairs, is correct-er. Therefore, let me take a moment and give you the PROS and CONS.

The PROS of Neuproz pairs exchanging a NEGATIVE energetic quanta:

1. We exist in a NEGATIVE branch of the universe and the exchange of NEGATIVE energetic quanta will conserve more energy, i.e. LESS degradation in intensely NEGATIVE environments.
2. We exist in a NEGATIVE branch of the universe and we use NEGATIVE magnetic energetic quanta to ISOLATE & DETECT

energetic quanta, i.e. positive energetic quanta are easier to isolate and detect.

3. The size of POROUS atomic particles is based upon the intercalation of NEGATIVE thermal energetic quanta, which would be detrimental to the exchange of 'positive' energetic quanta within the atomic nucleus.

4. The exchange of NEGATIVE energetic quanta about the atomic nucleus will LESSEN the concentration of NEGATIVE thermal energetic quanta **within** the atomic nucleus, which will ensure that the HARMONICS of energy exchange about the Neuproz pairs within the atomic nucleus are NOT bothered by average NEGATIVE thermal conditions...whatever that means. Also, less negative thermal energetic quanta within the center of the atomic nucleus will mean electrons will degrade LESS when undergoing Quanta Charging.

5. It is EXTREMELY unlikely that Quanta Charging NEGATIVE electrons will collide with NEGATIVE energetic quanta being exchanged by Neuproz pairs within the atomic nucleus.

6. The spherical nature of 'S' atomic orbitals makes a little more sense with regards to the exchange of NEGATIVE energetic quanta about the atomic nucleus, i.e. the APPEARANCE of positive energy moving about the circumference of the atomic nucleus.

7. The stimulated degradation of Neuproz particles results in the release of negative energetic quanta, which have the propensity

to NON-detrimentally modulate the negative energy moving about the center of the atomic nucleus, which determines atomic orbital arrangement, i.e. atomic orbital modulation via conservation of energy.

8. The exchange of a NEGATIVE energetic quanta between Neuproz pairs allows for fusion in really-REALLY hot negative environments, although it is energetically silly.

9. The exchange of a NEGATIVE energetic quanta between Neuproz pairs EXPLAINS why there is a LACK of mass variance between protons and neutrons...in this negative branch of the universe.

CONS of a **negative** quanta being exchanged about Neuproz pairs:

1. A major blow to my Ego.

2. A NEGATIVE energetic quanta being exchanged by Neuproz pairs has NOT been discovered.

Even though my working hypothesis is NOW that Neuproz pairs are exchanging a NEGATIVE energetic quanta, there is still a weird question rattling around my head: Was it easier to convince people of a highly ordered universe with the exchange of a 'positron' between Neuproz pairs? (The answer is: It doesn't matter because NOBODY has read my books.) In any event, with a new subdivision imparted to the theoretical environment within my head, I should take the time to NAME this new subdivision.   Hmmm, what should I call this theoretical NEGATIVE energetic quanta that is being exchanged between Neuproz Pairs? How about Neuprotron? (I was going to say Negatron, but Word Dictionary was under the assumption that that was already a word. Therefore, I went with the Neuprotron because now I get to add it to Word Dictionary…Yeah! It's the little things in life☺)

Now that I've added Neuprotron to my Word Dictionary, what does this mean for all my other postulates? Well, not much. It is still the same old story of a universe based upon ENERGY CONSERVATION…except with a NEGATIVE Neuprotron being exchanged between Neuproz pairs.

In conclusion, my Ego has taken yet another massive blow. But on the bright side, I got to add another Word to the Dictionary: **Neuprotron** – the negative energetic quanta that is being exchanged between Neuproz Pairs. All of which, makes a lot more sense with regards to the NEGATIVE branch of the universe in which we reside…theoretically. (On the off chance we accidently labeled positive energetic quanta with a 'negative' title, then I reserve the right to rearrange my nomenclature…regardless of what either of those terms actually mean.)

Chapter 7: Salty

With the freshly minted idea of Neuproz pairs exchanging a NEGATIVE Neuprotron within atomic nuclei, it is the optimal time to return to reality, i.e. introduce an ODD fact. Did you know that there is MORE **negative** chloride in the ocean than **positive** sodium? I mean, what are the odds of NEGATIVE energetic quanta existing LONGER within Earth's NEGATIVE magnetosphere within a negative branch of the universe? (That was sarcasm.)

In as much as I've postulated that the universe is based upon energy conservation via the forging of heavier matter **from** lighter matter and heavier matter **into** lighter matter, i.e. STEM matter selectively degrading into lighter matter based upon the external energetic environment, I NEED to reiterate that nothing lasts forever. You die, planets die, galaxies die, and matter dies. BUTT, what is MAJOR factor of matter death? Do positive elements degrade faster in a negative branch of the universe? Well, in all honesty, the answer is YES because protons are ALWAYS degrading negative magnetic energetic quanta to facilitate ANISTROPY and electrons have to be 'bathed' in negative energy to get them to degrade and release light. Therefore, from an evolutionary standpoint, the elemental composition of planets will evolve over time, i.e. positive elements will decay faster than negative elements. Unfortunately, based upon the postulate that STEM matter decays into lighter elements, it is theoretically possible that Earth was 'birthed' with more negative chloride and this was accentuated by positive elements decaying faster.

In the event that you don't believe or understand evolution, let me take a moment and explain it in terms you might understand. There are TWO birds with two DIFFERENT beaks. One bird, as a result of his 'weird' beak, can open up and eat five nuts a day. The other bird, with a traditional beak, can only open up and eat two nuts a day. As a result of the first bird EATING more, it has more energy to build nests, impress lady birds, and have sex. And the second bird, is LETHARGIC because it's not getting enough food. Now, multiply this 'food' difference over a hundred thousand years and it should be APPARENT that the bird with a 'weird' beak that eats five nuts a day will have MORE children. (FYI, even if you believe the Earth is only 6000 years old, birds have a very short life span, which allows for 'evolution' based upon environment, i.e. that's 6000 years of bird generations!) Or in human terms, one person gets one hamburger a day and the second person gets four hamburgers a day. Which one is going to have the energy to 'fuck' after a hard day's work? Hopefully, this clears things up for anyone that might have missed the chance to read Darwin's books. (They were boring and tough to read because English-speak evolves too...Lingua Franca!)

With the 'general' theory of energy over vast amounts of time firmly secured in your neurological reservoir, let's juxtapose this with the theory of energy over very short periods of time. Or in more explicit terms, salt and ice exists at -5° Celsius? (FYI, ice exists at 0° Celsius.) Unfortunately, before I get to a postulate with regards to salt and ice existing at -5° Celsius, I need to return to the macro-world to explain open and closed systems.

Mathematicians base all their calculations upon the ASSUMPTION that the universe is a CLOSED system, i.e. energy can be neither be created or destroyed. But, as per *Quanta Dynamics*, I postulate the universe is an OPEN system, which is based upon energy conservation because energy DEGRADES. For example, I postulated that Earth is an energetically OPEN system that is constantly degrading and diffusing negative energy. Therefore, to truly understand salt and ice existing at -5° Celsius, you need to remember the universe is an OPEN system. But, you also have to remember Atomic Thermal Capacity as well.

In the event that I postulate that the combination of salt and ice results in a GREATER degradation of negative thermal energetic quanta, i.e. -5° Celsius, then I have to deal with my PREVIOUS postulate with regards to stimulated degradation resulting in a GREATER negative energy release, i.e. ONE thermal energetic quanta will cause the release of ONE magnetic energetic quanta, which has MORE negative energy.

In the event that I postulate that the Atomic Thermal Capacity of POSITIVE sodium decreases the amount of **diffusing** NEGATIVE thermal energetic quanta from transitioning ice at 0° Celsius, then I have to compensate for this weirdness by FINDING the lost negative thermal energetic quanta being absorbed by the POSITIVE sodium.

As you can see, I'm in a bit of a predicament here. On one hand, stimulated degradation results in more energy. On the other hand, Atomic Thermal Capacity is mostly containment. All of which, leaves me with the last option: OPEN energetic system, which is an even tougher sell…even for me. Therefore, let me try and recalibrate all of this complexity back to *Quanta Dynamics*.

The fundamental conundrum around -5° Celsius is as follows: **Detection**. Or in scientific terms, the ENVIRONMENT of -5° Celsius results in the stimulated degradation of energy that is NOT detected thermally, which allows for an OPEN system at a cooler temperature. Or in simpler terms, positive sodium plus negative chloride and positive protons, results in the enhanced degradation of positive protons. And since we ALREADY know that protons release copious amounts of negative magnetic energetic quanta, i.e. Anisotropy, then we have a METHOD for the NON-detectable diffusion of negative energy away from salt and ice at -5° Celsius. Or in the simplest terms, negative magnetic energetic quanta have a TON of momentum in comparison to negative thermal energetic quanta, which means they diffuse quickly to be absorbed and destroyed by Earth's negative magnetosphere outside the water. Or in even weirder terms, the sodium/chloride/ice-water atomic orbitals results in an environment of atomic orbitals that do NOT degrade certain negative magnetic energetic quanta, which diffuse into the external magnetic environment without detection.

In conclusion, it took me a long time to figure out this METHOD. (You should have seen some of the crazy shit I came up with prior to this method.) And the method, just in case you missed it, is based upon SPECIFIC atomic orbitals degrading SPECIFIC energetic quanta, or NOT, and the diffusion of NON-detectable negative magnetic energetic quanta created by the chelation of transitioning ice to water by positive sodium and negative chloride. (FYI, this release of magnetic energetic quanta is NOT detectable because it is NOT bimodal, i.e. NO north or south.) All of which, gives a logical method per the basis of Quanta Dynamics: The universe is OPEN system; Energy degrades; Energy diffuses based upon the environmental constraints. And finally, evolution is real...deal with it.

Chapter 8: Justified

I postulate that Neutrons have more energy, but they have about the same MASS as protons. (I got bored with introductions.) First and foremost, protons are positive and are continually degrading in this negative branch of the universe. Secondly, and more importantly, the exchange of a Neuprotron, which is negative, adds mass to protons to make them neutrons. Unfortunately, to truly understand the logic of this event, we must re-INVISION the concept of fusion. Therefore, let's start with a quick review of Neuproz pairs via my working hypothesis.

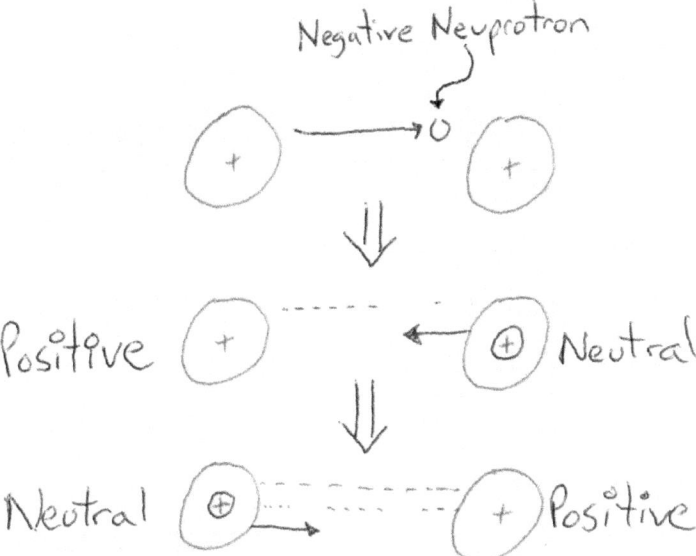

Figure 11: Neuprotron Relativity

As you can see in this figure, the exchange of a NEGATIVE Neuprotron between "positive" protons results in the formation of a proton-neutron complex. Unfortunately, the current theory of fusion complicates things.

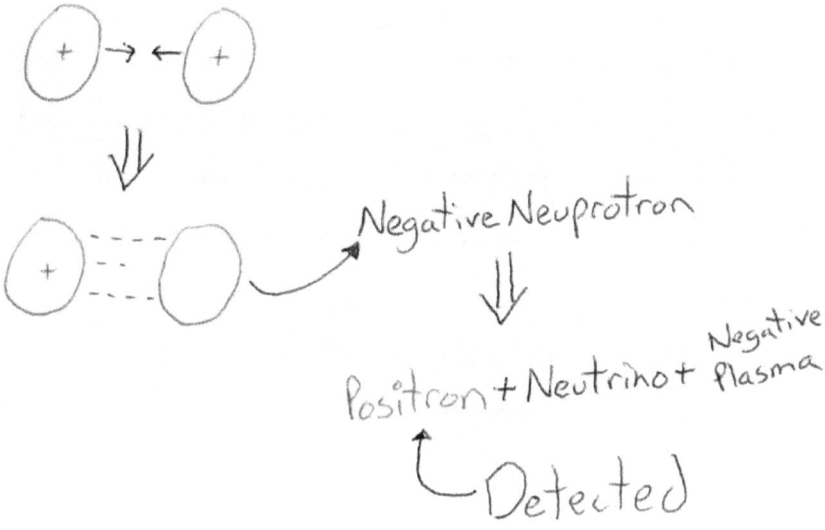

Figure 12: Detection

As you can see in this figure, the fusion of two positive protons is THOUGHT to cause ONE proton to release a Negative Neuprotron, which degrades to yield a positron, neutrino, and more than likely, some negative plasma. Or in simpler terms, the EXTREME negative magnetic energetic environment used to detect the "products" of fusion, results in the degradation of the NEGATIVE Neuprotron into negative plasma, neutrino, and positron, which is detected because it is positive. But, since protons are positive, we exist in a negative branch of the universe, and Quanta Dynamic's main premise is the continual degradation of energy, MORE THAN likely, the fusion of two NON-identical positive protons results in the release of energy from BOTH protons to form a Neuproz Pair.

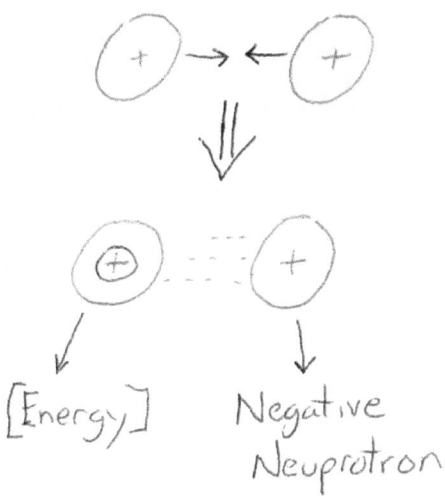

[Energy]   Negative
               Neuprotron

Figure 13: Quanta Dynamic Fusion

As you can see in this figure, the alignment of two NON-identical protons results in the release of [energy] from one proton and the release of a negative Neuprotron from the other proton to produce a Neuproz pair. With all that in mind, let's take a moment and review some very important concepts.

First and most importantly, Einstein's Equilibrium determines the energy within protons and neutrons. Second, the 'mass' of an object is based upon the VISIBLE positivity within said electronic structure. Third and finally, ONLY protons undergo fusion...damn it. I guess I have to make another correction.

Previously, I postulated that Neutrons ONLY create isotopes via intercalation. But, with this new postulate, the conservation of energy dictates that Neutrons and Protons can undergo fission to create a NEW Neuproz pairs BECAUSE this event would only result in the release of [energy] **instead of** [energy] and a Negative Neuprotron, per the logic of Figure 13. Granted, there will probably be years and years of discussion with regards to the 'fusion-ability' of between spectrums of Neutrons and Protons, but, from where I'm sitting, the fusion of Neutrons and Protons releases LESS energy, i.e. it conserves more energy.

In conclusion, it is REALLY difficult to detect very small negative energetic quanta within a negative branch of the universe using negative magnetic energetic quanta to focus and detect negative ENERGY. Therefore, our view of fusionary events is somewhat incomplete in that the positron and neutrino being released during fusion probably resulted from the degradation of a negative Neuprotron. Next, based upon a spectrum of continually degrading protons, more than likely BOTH protons undergoing fission are releasing energy. Also, the fusion of Neutrons and Protons conserves more energy. And finally, even though mass is a function of positivity within this branch of the universe, Neutrons contain more "energy" than protons, even though we can't **currently** quantify it.

Chapter 9: Esoteric

It is a tremendously weird thing to see ENERGY spurt from the Sun and then CURVE back towards the NEGATIVELY diffusive Sun, i.e. Sun Flares. For starters, the NEGATIVE Sun is continually diffusing NEGATIVE energy. Next, the behavior of Solar Flares is an unnerving event as it pertains to the AGE of stars. And finally, ALL of this gives a BEAUTIFUL mental picture of atomic particles.

I would normally start off by rendering an attack towards the Big Bang theory, butt, I think we're all beyond that. Therefore, the decomposition of energy and the release of it via solar flares will be dependent on the Star's composition. If the NEGATIVE solar flare curves back towards the star, then the star is probably OLD and contains a ton of POSITIVE protons, i.e. relatively positive. If the NEGATIVE solar flare does NOT curve back towards the star, then the star is probably YOUNG and EXTREMELY negative. All of which, is based upon relativity of 'charge' with regards to our 'negative' branch of the universe. (I would normally add in a quip about humanity working together blah blah blah, but let's be honest with each other: Nazis are what Nazis do...Just enjoy your miserably short life.)

With that tremendous visual impression impregnated into your mind, please visit NASA's homepage if it is not, imagine a positive proton is a Star, i.e. continually degrading based upon environment until its eventual demise. Now, the major difference between positive protons and stars is that the protons must be stimulated to degrade in order to DETECT their Flares, i.e. the release of a photon of light or a magnetic energetic quanta. Or in simpler terms, we haven't developed the technology to detect the energy that protons are continuously degrading. In any event, let's take a moment and imagine the UNIMAGINABLE timeline with regards to Proton Flares.

As odd as the image on the front of this book might seem, I figure I should describe it. In the essence of IMPERFECT PERFECTION and the degradation of God's Pet Singularity, each branch of the universe is imparted with UNIQUE tiny energy fragments. For the sake of clarity and nomenclature, let's call these tiny fragments Thermal Energetic Quanta. In our branch of the universe, thermal energetic quanta are negative, which results in the EXPANSION of positive matter. The more negative thermal energetic quanta there are around positive matter, the more massive the positive matter will be. All of which, thankfully, is modulated by the continual expansion and diffusion of the universe. Or in simpler terms, matter wouldn't last very long without diffusion.

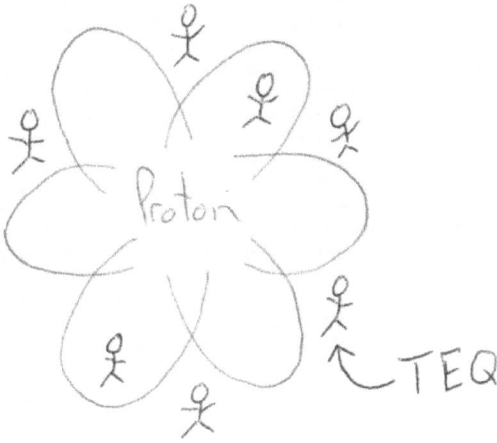

Figure 14: Simplicity

As you can see in this figure, all the tiny people represent negative thermal energetic quanta. And, the swooping lines represents the trajectory of the energy within the proton innervating with the environment. Normally, just like atomic orbitals, the energy within the proton is pushing away the negative thermal energetic quanta. But, as a result of the proton having an abundance of visible positive components, a number of tiny negative thermal energetic quanta people get caught up in 'swooping' nature of the energetic trajectory within the positive proton, which is dependent on the density and type of tiny negative thermal energetic quanta people.

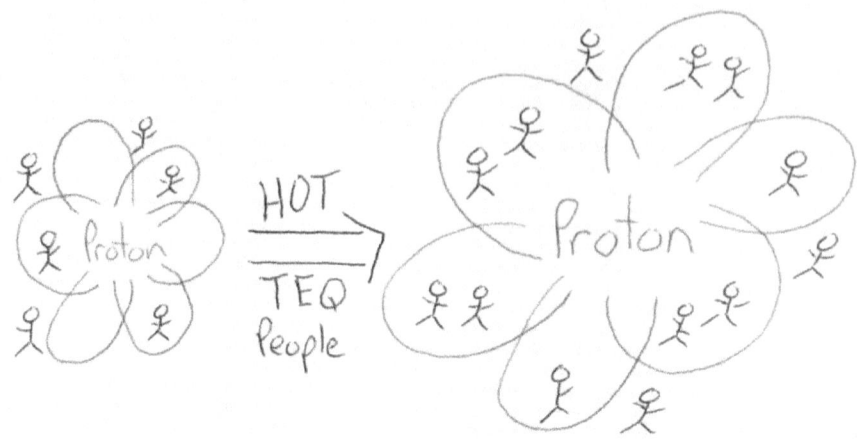

Figure 15: Expansion

Simply based upon columbic attraction, positive protons expand when they are exposed to more tiny negative thermal energetic quanta people, as described in the figure above. But, what HAPPENS when this Deathstar Proton is attacked?

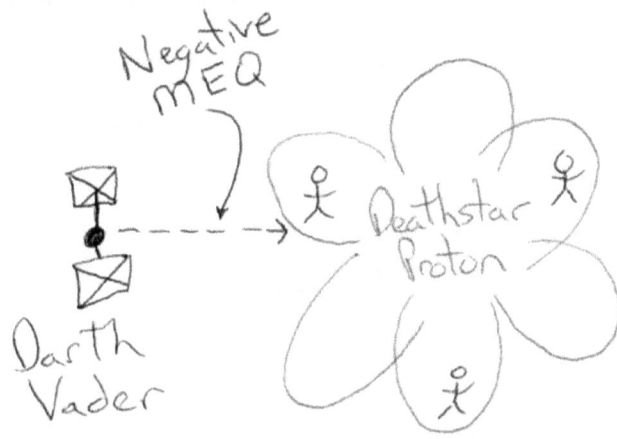

Figure 16: Use the Force of Magnetic Energetic Quanta

For reasons that should be abundantly clear to anyone that has read any of my books, I have drawn Darth Vader's tiny ship firing a NEGATIVE magnetic energetic quanta at the Death Star, i.e. my drawing sucks and the Imperial Fighters are really tough to draw. In the event that the NEGATIVE magnetic energetic quanta strikes a POSITIVE component of the positive Deathstar Proton, a TON of innocent/tiny negative thermal energetic quanta people are going to DIE and/or be BLOWN out into deep-dark-COLD positive space.

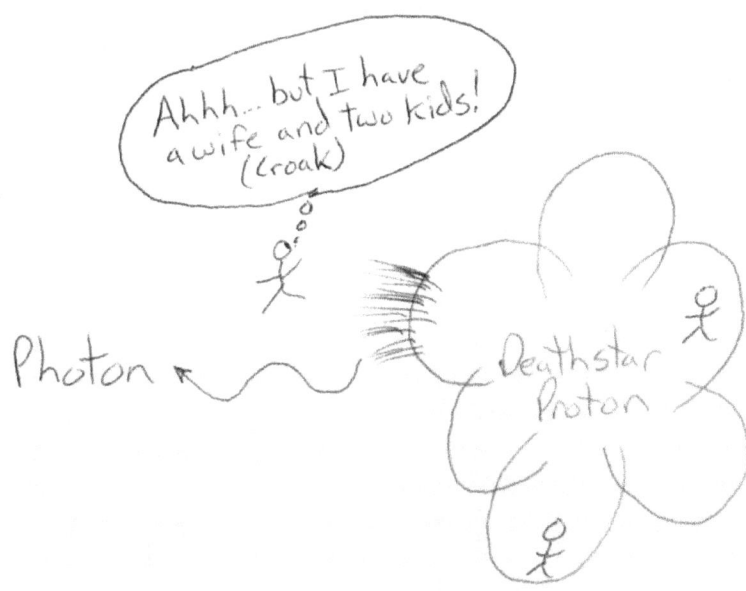

Figure 17: Fire in Compartment C

As you can see, the positive Deathstar Proton has NOT been destroyed. In fact, other than the proton releasing a TINY Proton Flare, i.e. some plasma and photon of light, the Deathstar Proton is unscathed. Unfortunately, as a result of the extreme number of tiny negative thermal energetic quanta people being BLOWN into deep-dark-COLD positive space, a formal investigation had to be held. The findings are as follows: *Quanta Dynamics*.

In conclusion, in as much as this chapter has been esoteric in nature, i.e. nature is esoteric, it is amazing the HARMONICS that exists within a universe based upon highly ordered energy and energy conservation. I would dare say that it is beautiful, but public opinion will probably disagree. Therefore, let me simply conclude by saying: Momentum towards ANYTHING cost lives.

Chapter 10: Stimulated Tooth Emission

Even though we would love to romance the idea that Darth Vader's pathetic shot across the bow of the Deathstar Proton would somehow destroy the WHOLE Deathstar Proton, it's not very scientific. What is scientific, is to wonder about the **method** behind the 'stimulated' emission of ONE measly photon and ONE poor Deathstar Proton TEQ worker, which died a horrible death in deep-dark-COLD positive space. Or in simpler terms, let's investigate the COMPLEXITY of the Deathstar Proton, which resulted in the release of ONE measly photon. (I realize that the 'Magic Bullet' theory was used in one of the Star Wars movies, which was VERY entertaining, but you know, Star Wars not scientific.)

With that delightful introduction out of the way, this chapter is going to be rather short because I lack the ability to draw. I mean, if I was M. C. Escher, then I would draw a 3-dimensional sphere of inner working gears, which would represent the synergistic movement of positive and negative energy within and about the Deathstar Proton. All of which, would be dynamic enough to allow the movement of 'gear fragments' to render new arrangements based upon the stimulated emission of ONE measly photon. But, I am not M. C. Escher. Therefore I will have to beg someone else to render this idea upon your imaginary event horizon. I guess you could watch that 'Zero Theorem' movie and juxtapose the complexity of a continually morphing problem to the energy within protons, but the movie wasn't that good. In any event, you can spend your time more wisely by finishing my book, going to a local café, having a croissant, and picking your nose in public. Unfortunately, I get the overwhelming feeling that I NEED to explain the 'gear' metaphor as it pertains to stimulated emission. Therefore, please feast your eyes upon my HORRIBLE drawing.

Figure 18: Columbic Dynamics 101

As you can see in this figure, Darth Vader has shot ONE tooth from a 'fictional gear', which has result in all the other teeth SPREADING out about the circumference of the 'fictional gear'. Granted, the loss of ONE tooth could result in a circumference change, but that is beside the point. The point is: The VARIATION in the 'fictional gear' will result in the modulation of ANOTHER 'fictional gear', which will eventually result in the emission of ONE photon. Or in scientific terms, the columbic repulsion of the resulting three energetic-teeth will result in their dispersion about the circumference of the 'fictional gear', which will STIMULATED another 'fictional gear' to release ONE photon tooth.

In conclusion, the Deathstar Proton is immensely complex AND dynamic. Therefore, when one gear loses a very tiny tooth, another gear compensates by emitting a NEGATIVE photon, which CAN cause the rearrangement of the whole dynamic gear system, i.e. 'gear fragments' being exchanged to create new gears and arrays of gears. And even though this METHOD is very vague and imaginative, it is more scientific than Star Wars...hopefully.

Chapter 11: Einsteinium

You know you've made an impact on humanity when they name a radioactive element after you. Unfortunately, I think Captain Einstein was dead when that happened, but that is beside the point. The point is: Einstein still exists in solid state...whenever a scientist gets the inclining to synthesize him. (That was a public education metaphor.) All of which makes me wonder: Can solid state NMR irradiation of some isotopes avail different catalytic properties?

Previously, per my unique intellectual genre, I postulated that electrons are stimulated to degrade within Earth's core to yield negative magnetic energetic quanta, which gives the Earth a protective NEGATIVE magnetosphere. As per the reason that NEGATIVE electrons are degrading to produce Earth's NEGATIVE magnetosphere? Well, we've got a little extra negativity to spare since we live in a NEGATIVE branch of the universe. Having said that though, protons and neutrons within the atomic nucleus degrade and release NEGATIVE magnetic energetic quanta, albeit less vigorously as a result of the protective layer of NEGATIVE electrons. All of which, results in a method for the modulation of NEGATIVE Neuprotrons between Neuproz pairs, which forges the directionally dependent electron Quantum Charging, i.e. atomic orbitals, in lighter elements.

With that pleasant postulative review in mind, a METHOD to magnetically responsive elements pops into mind. If it doesn't automatically jump into your mind, let me be a synaptic-friend and help you with a figure.

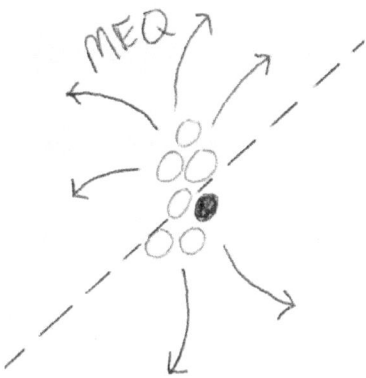

Figure 19: UN-symmetric C-13 Atomic Nucleus

As you might be able to ascertain from this figure, the EXTRA blue Neutron results in the UN-symmetric release of Magnetic Energetic Quanta (MEQ) from carbon-13. (FYI, it looks kind of weird because I was trying to draw one side of a 2[3N] Neuproz Complex with an extra Neutron, i.e. C-13) Or in simpler terms, the atomic nucleus NOW has a Pinkie Tilt Factor. As a result of this Pinkie Tilt Factor, non-symmetric atomic nuclei can be detected within NMRs.

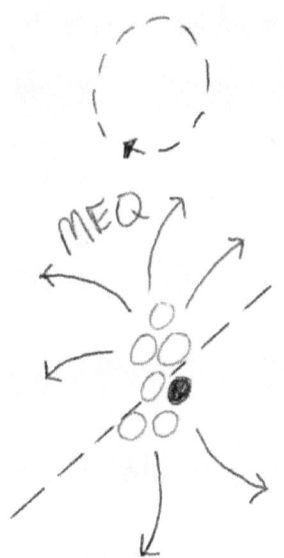

Figure 20: Procession Two-step

Now, I'll be honest with you all, this figure is kind of vague. What the hell is the dotted 'circle' and what is the procession two-step? Also, how on Earth does a HUGE-ASS atomic nuclei **process** like tiny protons? Thankfully, I have JUST enough education to answer those two poignant questions. First and foremost, the dotted 'circle' represents the restricted rotation as it pertains to a NMR's magnetic environment. Secondly, and more importantly, it takes a LOT of time for HUGE atomic nuclei to relax by procession...relative to tiny protons. And while we're on the topic, I might as well point out the fact that noise to signal ratio of carbon NMRs is much greater than proton NMRs because of the constraint of free rotation about carbon atoms, as it relates to detectable rarefactions in the negative milieu, as a result of the other things attached to the carbon atoms. (If I don't deserve a doctorate for that last sentence alone, then I have truly wasted $155,000 dollars and eleven years of my life.)

In any event, where does all this leave us? Well, let's see…What the hell was the point of this chapter? Ummm, catalysis? (Yeah, I'm pretty sure it was catalysis.) Wait, magnetically enhanced catalysis…that sounds better. Unfortunately, this means we need to review a few more things before we get to the juicy part of this streaming cognition. For starters, atomic nuclei can release either (R) or (S) Magnetic Energetic Quanta…as per the figure below.

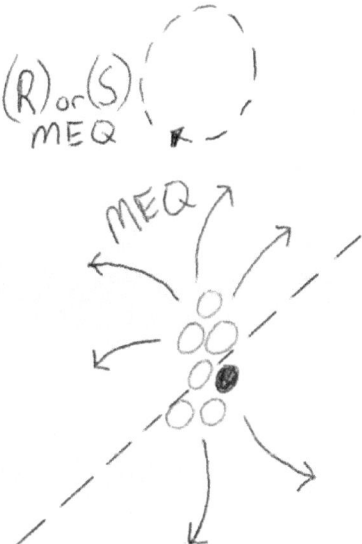

Figure 21: Figgy

What you can NOT see in this figure is: There are TWO possible isotopes, which are NOT 'enantiomeric'. That is NOT to say that Neutrons that are intercalated in the Neuproz complex DIFFERENTLY will result in DIFFERENT isotopes, but I am restrained by the State Department from talking about all that based upon National Security. (FYI, based upon the size and the number of Neuproz groupings within an element, the number of DIFFERENT isotopes varies.) Therefore, let's just imagine we can ISOLATE one isotope and directionally secure it to the surface of a solid. Unfortunately, this is where things get theoretically-foggy.

In as much as the degradation of 'matter' within the atomic nuclei is ULTIMATELY based upon the perfusion of negative energy into and around the atomic nucleus, the SELECTIVE degradation of the 'matter' will be an exciting new scientific field in the future. Therefore, let's just simply imagine that we can directionally modulate the release NON-symmetric magnetic energetic quanta with directionally SPECFIC radio wave bursts, i.e. rarefactions in the energetic milieu.

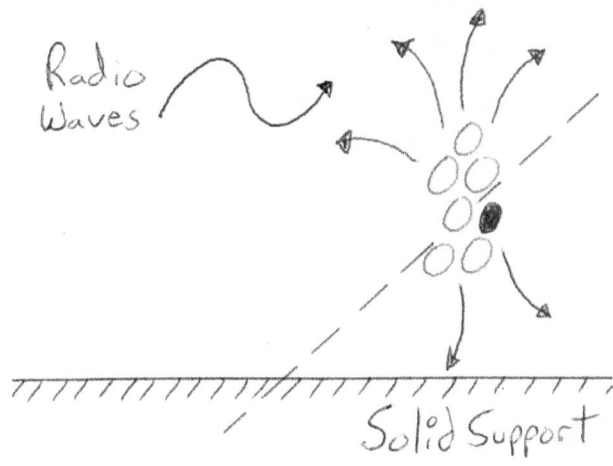

Figure 22: Solid Support

As you can see in this figure, the CONSTANT vibration of directionally specific radio waves towards an atom attached to a solid support will MODULATE the MEQ arrows, which will modulate the placement of atomic orbitals, which will modulate reactivity. Therefore, once the OPTIMIUM direction of radio wave irradiation has been determined per desired reactivity of an atomic orbital, magnetically enhanced catalysis will be profitable.

For the keen observer and thinker, this selective modulation of atomic orbitals based upon environmental factors is an explosion in the head. If you could directionally irradiate the solid support with ONE magnetic energetic quanta, would it have the same effect? (Similar to the METHOD behind why water is denser at 4° Celsius) If you add a solid supported magnetically enhanced catalysis to a chiral NMR, could you do some crazy abnormal things? (Fun question, right?!) If you could make a crazy chiral NMR environment that stabilized radioactive elements such that they could make useful and EXTRAORDINARY magnetically enhanced catalysis, could we reincarnated Einstein to save the world? (Frankenstein Einstein haunts my dreams.) If both of Earth's magnetic poles release a **different** magnetic energetic quanta, then would catalysts within NMRs behave differently at OPPOSITE poles? (Weird, right?)

While I'm on the subject of elemental magnetism, I need to take a moment and clarify something: Magnetic energetic quanta released by elements behave differently within different magnetic environments. For example, negative magnetic energetic quanta have more 'energy/momentum' within Earth's negative magnetic energetic quanta magnetosphere because of the attraction of OPPOSITE magnetic energetic quanta. On the other hand, negative photons have

MORE 'energy/momentum' in deep-dark-COLD positive space because the expansive-destruction of 'more positive' negative magnetic energetic quanta. Or in simpler terms, Einstein's Equilibrium. Or in figurative terms, let's see at how a lighter element would behave within Earth's NEGATIVE magnetic energetic quanta magnetosphere and in deep-dark-COLD positive space.

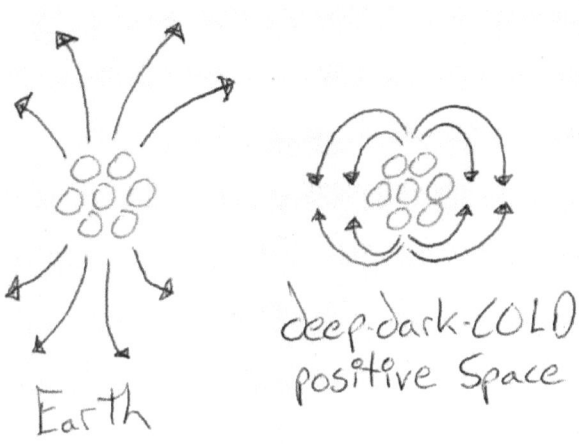

Figure 23: Two Drastically Different Environments

As you can see, the magnetic energetic quanta being released by electrons and protons ARC outwards toward Earth's external magnetosphere. On the other hand, the magnetic energetic quanta being released by electrons and protons create a miniature magnetosphere around the lighter element when they are in deep-dark-COLD positive space. And even though deep-dark-COLD positive space results in less STIMULATED degradation, i.e. less negative thermal energetic quanta, the MOVEMENT of the negative magnetic energetic quanta ARE a FACTOR in the stability of elements per their environment. All of which, brings us to a mind-bottling postulate. (No figures, sorry.)

In as much as the ARC of negative magnetic energetic quanta of lighter elements, within a larger magnetosphere, results in the placement of atomic orbitals based upon the **columbic repulsion** between negative electrons and negative magnetic energetic quanta, the COMBINATION of negative magnetic energetic quanta to form plasma is the basis of atomic orbital placement in deep-dark-COLD positive space. Or in simpler terms, deep-dark-COLD positive space is deep-dark-COLD and POSITIVE. And as a result of being in a negative branch of the universe and INTER-shell dynamics being based upon the irradiation of negative energy outwards, which is based upon stimulated emission, the atomic orbitals arrange near the COLLISION of the negative magnetic energetic quanta in deep-dark-COLD positive space. Or in the simplest terms, conservation of energy WITHIN a NEGATIVE magnetosphere will place electrons AWAY from the collision of negative magnetic energetic quanta. But do to POSITIVE nature of deep-dark-COLD positive space, i.e. the conservation of energy OUTSIDE protective magnetospheres, electrons swirl AROUND the collision of negative magnetic energetic quanta. ALL OF WHICH, is DIFFERENT for heavier elements. As a result of the MASSIVE amount of non-Quanta Charging negative electrons crowding around heavier elements, there is a different rate of thermal energetic quanta diffusion, which means the heavier elements do NOT conserve energy by placing electrons AROUND the collision of negative magnetic energetic quanta in deep-dark-COLD positive space. But, when heavier elements are placed within strong magnetic environments, the pattern of magnetic energetic quanta being released changes and the amount of 'internal' negative magnetic energetic quanta being created distorts the placement of the atomic orbitals just enough to destabilize the heavier element, i.e. shorter life span. That is not to say that ALL heavier elements experience this lifespan modulation based upon environment, but beyond a certain point in the periodic table, all heavier elements become radioactive...at least within our negatively diffusive Solar environment. (FYI, I'm probably going to devote a whole book to this topic, so please don't blow a blood vessel trying to understand it...and all the minutia.)

In conclusion, even though the magnetic energetic quanta being released by atomic nuclei don't make a magnetosphere-shell similar to Earth when within Earth's magnetosphere, the negative magnetic energetic quanta do a good job of forging atomic orbitals via modulating the movement of NEGATIVE Neuprotrons within the atomic nucleus and electrons outside the atomic nucleus. All of which, results in a Pinkie Tilt Factor when the atomic nucleus is imparted with a cordial neutron, which allows for SLOW procession and detection by NMR. And finally, environment is everything unless you can forge a new world in your head...Thanks Einstein!

Chapter 12: Cold Dinosaurs

A long time ago in a galaxy far far away, I had a heated argument with a right wing religious preacher: He was mocking carbon dating and I was trying to defend logic post World War II. Regardless of the fact that I lost the argument and he was sort-of-right with regards to carbon dating, his contribution to science will FOREVER infuriate him. (#☺) Quite simply, stem-matter degrades into lighter matter, i.e. carbon-14, which means carbon dating is ONLY a relative concept, i.e. NOT definitive. Or in simpler terms, the relative loss of carbon-14 in the environment is TOOL to determine the age of things, but in the PRESENCE of strong volcanic or asteroid activity, which will impart Earth with more carbon-14 and/or stem-matter, it becomes a relative tool. Unfortunately, right wing religious zealots will WARP this RELATIVE idea into whatever hidden agenda they require to avoid crushing depression every time they open their eyes or go to a museum. And even though I have more in common with scientists, I still have a bone to pick with them...mostly because they are the ones picking over the bones.

Based upon the working hypothesis within *Plinth*, the oxygen content of Earth was much greater during the time of dinosaurs, which was the main factor in the selection of larger animals. Or in simpler terms, each breath had greater energy, which allowed the animals to grow larger and larger. As a result of this high oxygen content, animals had to have protective scaly skin. After the comet that caused the flood that killed the dinosaurs, Earth's atmosphere EVOLVED to contain LESS oxygen, which resulted in the evolution of hair instead of scales. (Hair was beneficial for the new climate, which I'll talk about in a bit.) The important thing to remember is that the extraordinary growth of dinosaurs was the result of higher atmospheric oxygen, which REQUIRED the dinosaurs to have scales.

Now, the current belief is that dinosaurs existed in a luscious HOT tropical environment. But, if dinosaurs had scales as a result of an atmosphere with more oxygen, then dinosaurs did NOT SWEAT. Therefore, a warm blooded dinosaur with an INSANE metabolism, scales, and the inability to sweat, living in the HOT tropics, seems kind of fishy. Or in simpler terms, I think the time in which dinosaurs existed was a tundra-esk environment that ALWAYS had snow on the ground. All of which means, if a dinosaur died and wasn't eaten, then it froze. When the dinosaur eventually thawed, probably when the planet became tropical, then it APPEARED as if the dinosaurs died around the 'tropical' time period. Who knows, the dinosaurs might have died 100,000 years before the 'tropical' time period? Or in scientific terms, the asteroid dust that blocked out the Sun and eventually made the Earth freeze over, probably changed the composition of the Earth's atmosphere, i.e. decreased the oxygen content within the atmosphere. I guess the most poignant question to ask is: Do volcanic eruptions release a lot of nitrogen? The second most poignant question to ask is:

Does Nitrogen have a GREATER Atomic Thermal Capacity than Oxygen? Well, if the Nitrogen does have a GREATER Atomic Thermal Capacity, which I think it does simply on the fact that it normally exists in a positive state, then an increase in Nitrogen in the atmosphere will result in a warmer tropical environment because MORE water will diffuse into the atmosphere.

So let's recap. Dinosaur time was a snowy tundra because:

1. Increased atmospheric oxygen enabled the metabolisms needed to grow large quickly, i.e. dinosaurs are LARGE.
2. Increased atmospheric oxygen meant that dinosaurs had to have protective scales.
3. Scaly skin meant dinosaurs couldn't sweat, which means they probably lived in colder climates.
4. As a result of living in colder climates, frozen dead dinosaurs thawed during tropical times, which means they looked like they lived in tropical times.
5. The comet that caused multiple eruptions, resulted in more Nitrogen in the atmosphere, which increased the average atmospheric temperature.
6. As a result of the GREATER Atomic Thermal Capacity of Nitrogen, the atmosphere was warmer and imparted more water to the atmosphere, which increased the number of strong weather events that tossed unsuspecting, albeit stupid, animals into new environments to aid in Old Testament Evolution. (#Amen)

As for the reason why we **don't** have a fossil record of ALL the dinosaurs that ever existed is because MOST of the dinosaurs existed on a mostly frozen planet that melted and dumped their bodies into the oceans. All of which makes we wonder if dinosaurs didn't evolve from some whale heritage. I mean if you think about it, the statistical chance of a dinosaur's body being preserved perfectly frozen until the time when foliage was engulfing the Earth is pretty rare, i.e. dinosaur bones are pretty rare. Or in simpler terms, there are only very few deposits of dinosaurs, which occur in areas with a HIGH rate of sedimentation. Or in other words, the perfect mixture of conditions to SLOWLY thaw a frozen dinosaur and place him/her in a bed of sediment.

And while I'm on the topic of evolution, I have a chicken bone to pick with the general public. In as much as Evolution has been spouted as being the 'survival of the fittest', there are many examples of 'survival of the fittest GROUPS' in nature. For example, have you ever seen a rabbit run away? If you have, then you know the underside of their tail is white. Now, WHY would it be an evolutionary advantage for bunny to have a big BRIGHT white dot on their ASS whenever they are running away from a predator? Well, in terms of an INDIVIDUAL bunny, it is a SEVERE detriment. But in terms of a GROUP of bunnies, it is quite confusing to see all these bright white dots zig-zagging every which direction. Therefore, evolution has favored a detrimental INDIVIDUAL characteristic, but a favorable GROUP characteristic, i.e. there's micro-Darwinism and macro-Darwinism. All of which means, the next time someone uses micro-Darwinism to justify their selfishness, you can say

"Well from a macro-Darwinistic stand point, we have a better chance of bunny ass survival." (Then giggle to yourself.) Butt, just remember that we're all in this together, which means you'll probably have to tolerate that schlub or schlubette for a bit longer. Or in simpler terms, Fraud suggests you keep your mouth shut because trying to correct idiots usually ends badly for the less IDiotic, i.e. the person with LESS "inherited instinctive impulses of the individual as part of the unconscious," as per The Oxford Dictionary.

In conclusion, the relative carbon dating of frozen dinosaurs to the decayed vegetation surrounding the dinosaurs might be statistically-negligible because the comet that killed the dinosaurs caused Earth to spew out more 'carbon-14' into the air. Or in linear terms: Comet, dinosaurs freeze, Earth spews more 'carbon-14', millions of years go by, Earth evolves into a tropical paradise with about the same 'carbon-14' content, dinosaur icebergs melt, deposits the dinosaurs in a sedimentary tropical paradise, and carbon-14 content is very similar between the foliage and dinosaurs. All of which, is based upon dinosaurs having: Scales, which protected them from the massive oxygen content within the atmosphere but inhibited them from sweating; insane metabolisms, which allowed them to grow huge. (Have you ever wondered if T-rex could swim?) Or in rightwing religious terms, men and dinosaurs frolicked together in a veritable paradise when you factor out ALL known physiological parameters between the two entities.

Chapter 13: Garrulous

In as much as this book hasn't been garrulous, Einstein isn't an asymptote...or an idiot. Of course he did idiotic things, but that is nothing new to humanity. The truth is, on Humanity's Gaussian curve, Einstein was really-REALLY far away from the majority of humanity. As for the reason why he put himself there, that can only be speculated upon. Butt, as for the reason why I'm trying to push myself towards Einstein's relative region along Humanity's Gaussian curve, it is because there is some sort of connection between humanity's general populous and the distant thin regions in which Einstein, and his esteemed colleagues, existed. Maybe Freud or 2Pac had a word to describe this connection? In any event, trudging forward is just something humanity has been doing for centuries. Have we evolved? Most definitely. Do we need to keep evolving? Just look around at how we treat each other. Will this book make a difference? To me, yes. Therefore, let me take a moment and review some of the prominent things I've postulated over the course of the NON-garrulous book.

Environment is a big deal. So much so, that the longevity of elements are fashioned based upon the environment. In particular, lighter elements exist longer in regions of higher negativity in this negative branch of the universe and heavier elements exist longer in regions of lower negativity. Or in simpler terms, the general trend of half-lives we have come to embrace with regards to the periodic table is backwards in deep-dark-COLD positive space. Also, as a result of *Imperfect Perfection*, the smallest stable quanta forge the dynamics of larger energy fragments and their association, which forges unique elements. More specifically, the smallest stable energetic quanta plays a part in proton-neutron degradation, which modulates the placement of atomic orbitals directly and via Neuprotron trajectory within the atomic nucleus.

Even though there was much-much more in short book, especially with regards to my working hypothesis about peculiarities within the universe, there is no point in beleaguering you...especially if you made it this far. Therefore, please accept my humblest apologies for being an arrogant ass. I don't understand people and I especially don't understand why the world is the way it is, but I think it can be better. Best of luck and don't forget to say your prayers!

www.ingramcontent.com/pod-product-compliance
Lightning Source LLC
Chambersburg PA
CBHW070939180526
45168CB00003B/1102